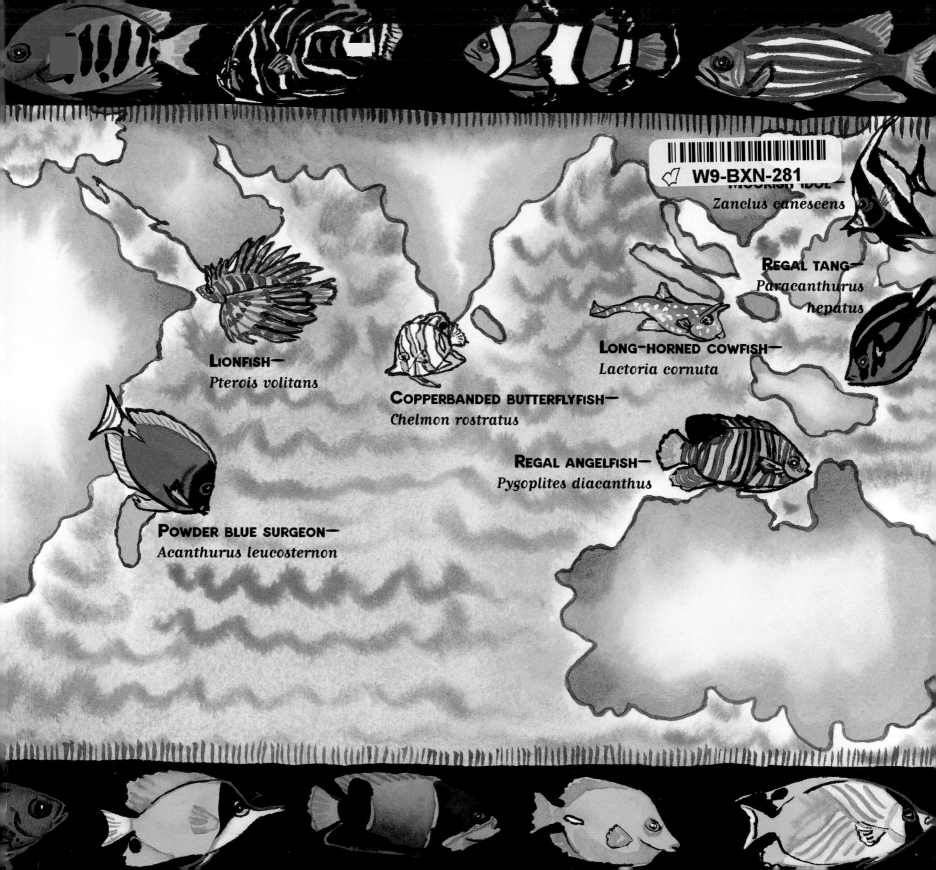

MOORISH IDOL—
Zanclus canescens

REGAL TANG—
Paracanthurus hepatus

LIONFISH—
Pterois volitans

LONG-HORNED COWFISH—
Lactoria cornuta

COPPERBANDED BUTTERFLYFISH—
Chelmon rostratus

REGAL ANGELFISH—
Pygoplites diacanthus

POWDER BLUE SURGEON—
Acanthurus leucosternon

Colorful Captivating
Coral Reefs

Dorothy Hinshaw Patent   Illustrations by **Kendahl Jan Jubb**

Walker & Company
New York

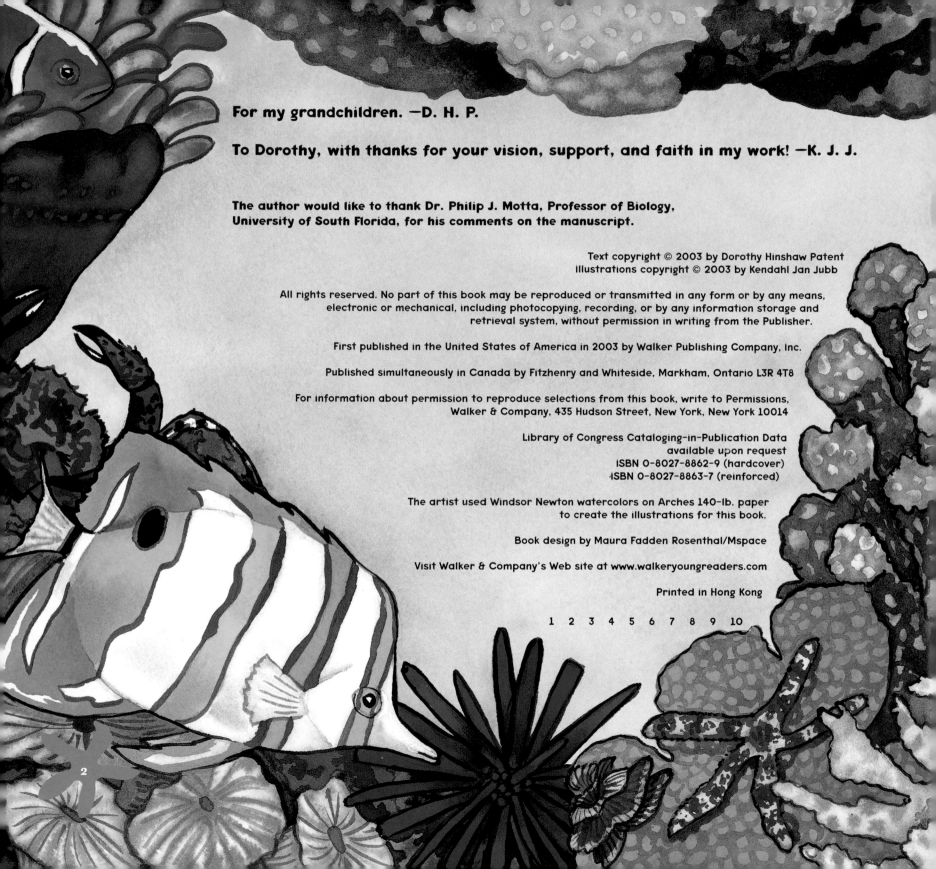

For my grandchildren. —D. H. P.

To Dorothy, with thanks for your vision, support, and faith in my work! —K. J. J.

The author would like to thank Dr. Philip J. Motta, Professor of Biology,
University of South Florida, for his comments on the manuscript.

Text copyright © 2003 by Dorothy Hinshaw Patent
Illustrations copyright © 2003 by Kendahl Jan Jubb

First published in the United States of America in 2003 by Walker Publishing Company, Inc.

Published simultaneously in Canada by Fitzhenry and Whiteside, Markham, Ontario L3R 4T8

For information about permission to reproduce selections from this book, write to Permissions,
Walker & Company, 435 Hudson Street, New York, New York 10014

Library of Congress Cataloging-in-Publication Data
available upon request
ISBN 0-8027-8862-9 (hardcover)
ISBN 0-8027-8863-7 (reinforced)

The artist used Windsor Newton watercolors on Arches 140-lb. paper
to create the illustrations for this book.

Book design by Maura Fadden Rosenthal/Mspace

Visit Walker & Company's Web site at www.walkeryoungreaders.com

Printed in Hong Kong

1 2 3 4 5 6 7 8 9 10

Coral reefs are among the most beautiful and busiest places on Earth. A coral reef is like a big city. Different varieties of coral form the "buildings." Colorful fish swim around through the spaces in and between the corals and their branches, hiding here, feeding there. Other animals, such as worms, sea urchins, sea stars, shrimps, and crabs, also live on coral reefs. Altogether, almost a quarter of all known species of marine life depend on coral reefs, including 700 kinds of coral and 4,000 species of fish.

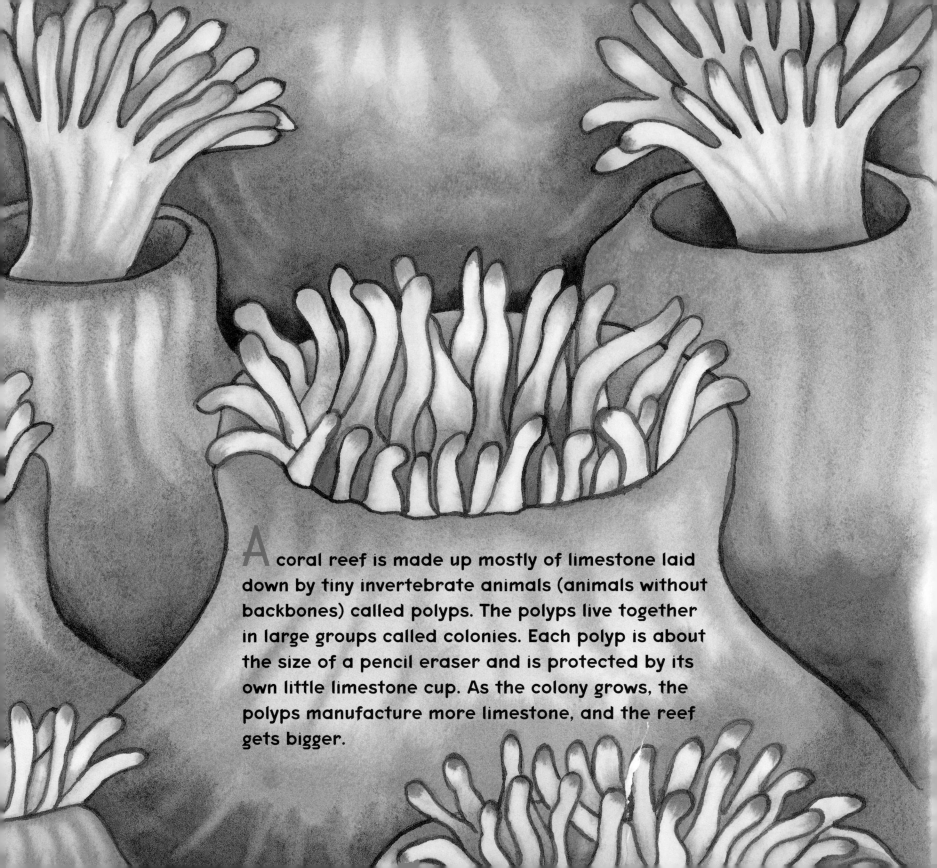

A coral reef is made up mostly of limestone laid down by tiny invertebrate animals (animals without backbones) called polyps. The polyps live together in large groups called colonies. Each polyp is about the size of a pencil eraser and is protected by its own little limestone cup. As the colony grows, the polyps manufacture more limestone, and the reef gets bigger.

LOBE

TABLE

ANTLER

FINGER

PLATE

CRUST

Corals that make reefs by laying down limestone are called hard corals. Coral reefs are found in clear tropical seas or nearby, where the water temperature stays above 65 degrees Fahrenheit. Reef-building corals thrive only as far below the surface as light can reach. They gradually disappear in water deeper than 100 feet.

5

There's a good reason why reef-building corals need sunshine. Each coral polyp has a central mouth surrounded by soft tentacles. When something touches a tentacle, special stinging cells are set off and attack.

6

Hard corals don't need to catch all their food, however. They get help from microscopic single-celled algae that live inside them. The algae harvest energy from the sunlight and transfer much of that energy to the coral. Most animals get all their energy from the food they eat, but the corals get up to 90 percent of their energy from this algae. In turn, the coral provides a safe environment for the algae.

LOBE

Hard corals come in many different shapes. Some, like lobe coral, make large mounds. Others are branched and look like antlers (elkhorn coral) or human fingers (finger coral).

Hard corals grow at different rates. Lobe coral grows less than an inch a year. Branching corals grow much faster, often more than four inches a year.

FINGER

ANTLER

9

Unlike reef-building corals, orange cup coral doesn't have algae in its cells, so it must catch its own food. It lives in shaded areas such as in caves and under overhangs. It can also live in waters deeper than 120 feet, where there is no light.

Other corals, called soft corals, don't live in limestone cups at all. Their colonies are flexible and take on many different shapes. Soft tree coral forms a beautiful big colony that looks like a feathery orange bush. Sea fans and sea whips are small soft corals that sometimes live on reefs, but they are not reef builders.

SEA WHIP

SEA FAN

11

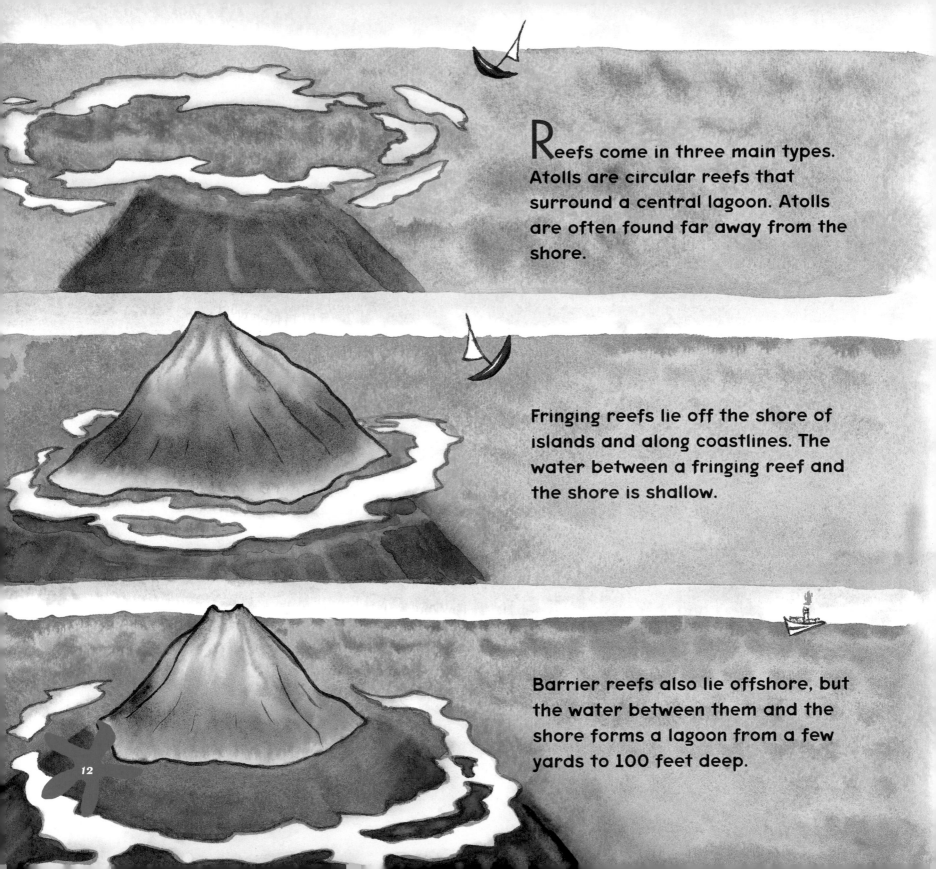

Reefs come in three main types. Atolls are circular reefs that surround a central lagoon. Atolls are often found far away from the shore.

Fringing reefs lie off the shore of islands and along coastlines. The water between a fringing reef and the shore is shallow.

Barrier reefs also lie offshore, but the water between them and the shore forms a lagoon from a few yards to 100 feet deep.

GREAT BARRIER REEF

The most famous reef is
Australia's Great Barrier Reef. This giant is
actually made up of more than 2,800 smaller reefs
that stretch along Australia's eastern coast. The Great
Barrier Reef is more than 1,200 miles long and is home to
340 kinds of coral and more than 2,200 species of fish.

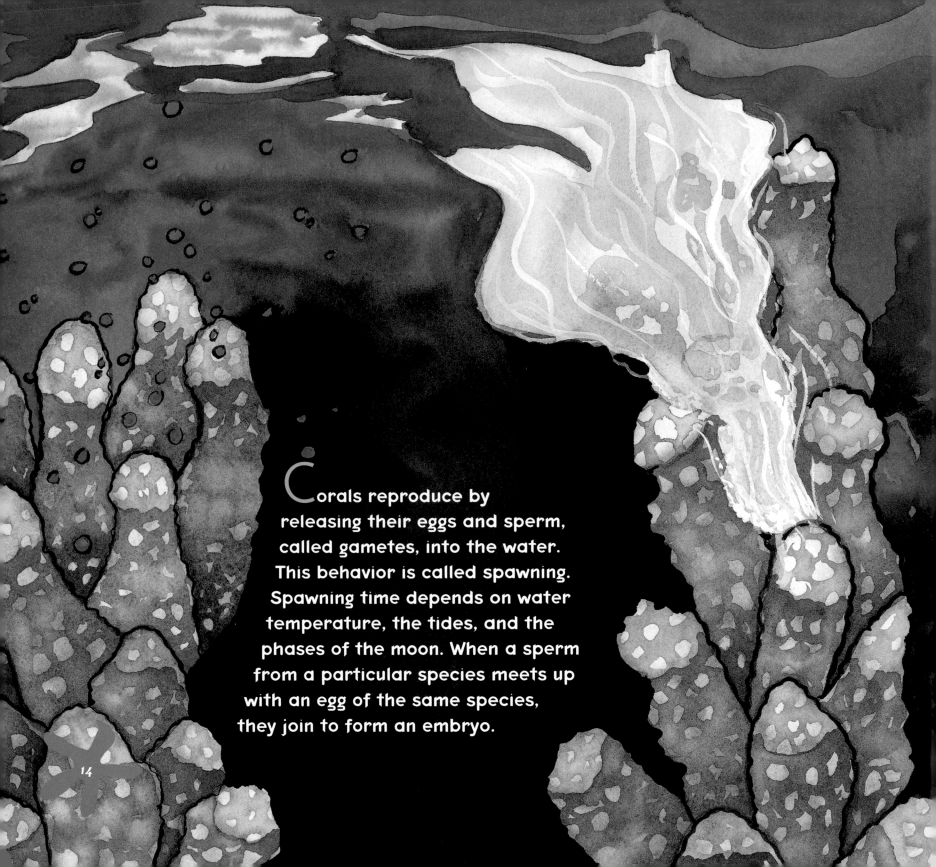

Corals reproduce by releasing their eggs and sperm, called gametes, into the water. This behavior is called spawning. Spawning time depends on water temperature, the tides, and the phases of the moon. When a sperm from a particular species meets up with an egg of the same species, they join to form an embryo.

This cell divides many times to form a tiny oval-shaped larva covered with tiny hairlike cilia that help it swim.

Coral larvae are attracted to light and swim toward the surface of the sea. There they live in the mixture of tiny plants and animals called plankton. Water currents carry the coral larvae away from the reefs they came from. Many larvae die, but after a few days to a couple of months, others find a good place to settle and grow into new coral colonies.

It's not surprising that the reef, with all its nooks, crannies, and branches, is home to many kinds of colorful fish. Their brilliant hues serve a number of purposes, from allowing males and females to recognize one another to providing camouflage among the colorful corals. Bright colors can also be a warning—don't bite me or you'll be sorry! The yellow tang, for example, has razor-sharp white spines by its tail to protect it. After one mistake, a predator knows to avoid this dangerous fish.

Color patterns can also confuse predators. It's hard to tell which end is which when the long-nosed butterflyfish swims along the reef. Dark bands across the eyes of some species can make their eyes hard to see. Many reef fishes, such as convict tangs, have bright black-and-white bands. A predator zeros in on the pattern, then the tang flees, its black-and-white stripes confusing the predator.

YELLOW TANG

CONVICT TANG

LONG-NOSED BUTTERFLY

NEEDLEFISH

Fish of all shapes live on reefs, too. Chunky boxfish, so well armored that only their mouths, eyes, and fins can move, awkwardly poke around the reef looking for food. Long, skinny needlefish float just below the water's surface.

BOXFISHES

Many reef fish are disk shaped. From the side, they are round or oval. But from the front or back, they are very slim and therefore hard to see.

18

Reef fish range in size from tiny to huge. Many boxfish, blennies, and other species never grow longer than five or six inches, whereas moray eels can reach eight feet in length. Giant potato cod, more than six feet long and weighing 200 pounds, also live along reefs. One place they gather is in a deep spot called Cod Hole on Australia's Great Barrier Reef.

POTATO COD

Reef fish feed in a variety of ways. Some fish, such as the leopard blenny of Hawaii, eat the coral itself. Other fish, like the threadfin butterflyfish, eat a more varied diet that includes sea anemones, worms, and small crustaceans.

THREADFIN BUTTERFLY

LEOPARD BLENNY

PARROTFISH

Parrotfish take bites from broken-off coral, but their food is actually the algae that grow within the coral. They are named for their strong, beaklike mouths. They bite off chunks of coral, then grind them using special plates at the backs of their throats. Their intestines use the algae in the coral as food, and the ground-up coral comes out in their feces as sand. You can thank parrotfish for much of the sand on the world's beaches!

21

The reef is also home to predators that feed on other animals. The Red Sea lionfish has a striped body and long, branching fins. It swims very slowly as it stalks its fishy prey. When it gets close enough, it rushes forward and engulfs the prey in its huge mouth.

The octopus is a predator that eats shrimps, crabs, mussels, and clams. But it can also be prey to such fish as moray eels, which hide in reef cavities waiting for a tasty fish or octopus to come close. The octopus can change its color pattern to match the background, making it hard for both prey and predators to see. If an octopus is threatened, it releases a cloud of ink. The ink not only hides the octopus, but can also blind the predator and confuse its sense of smell.

Reef fish also eat other foods. Some feed on plankton. Others eat larger kinds of algae, called seaweed. Damselfish that eat plankton hang out in groups above the reef, picking bits of food from the water. Those that eat seaweed keep to themselves. Each has its own territory along shallow parts of the reef where it feeds and chases away intruders.

Reef fish also eat invertebrate animals such as worms and clams. Goatfish are so named because the whiskerlike barbels around their mouths resemble the beard of a goat. Goatfish use the barbels to feel and taste the sandy bottom by the reef in search of food.

25

SHRIMP

SEA URCHIN

Fish and coral are the most obvious reef animals. But coral reefs are home to thousands of other kinds of animals, too. Every nook and cranny of the reef is home to such creatures as shrimps, crabs, and sea urchins.

Sea stars move slowly over the reef on thousands of tiny suckerlike feet, hunting for clams and mussels to eat, while fan worms glue their protective tubes to the reef. They extend their beautiful fans to feed on plankton. If danger threatens, fan worms can pull back into their tubes with lightning speed.

FAN WORM

CRAB

SEA STAR

27

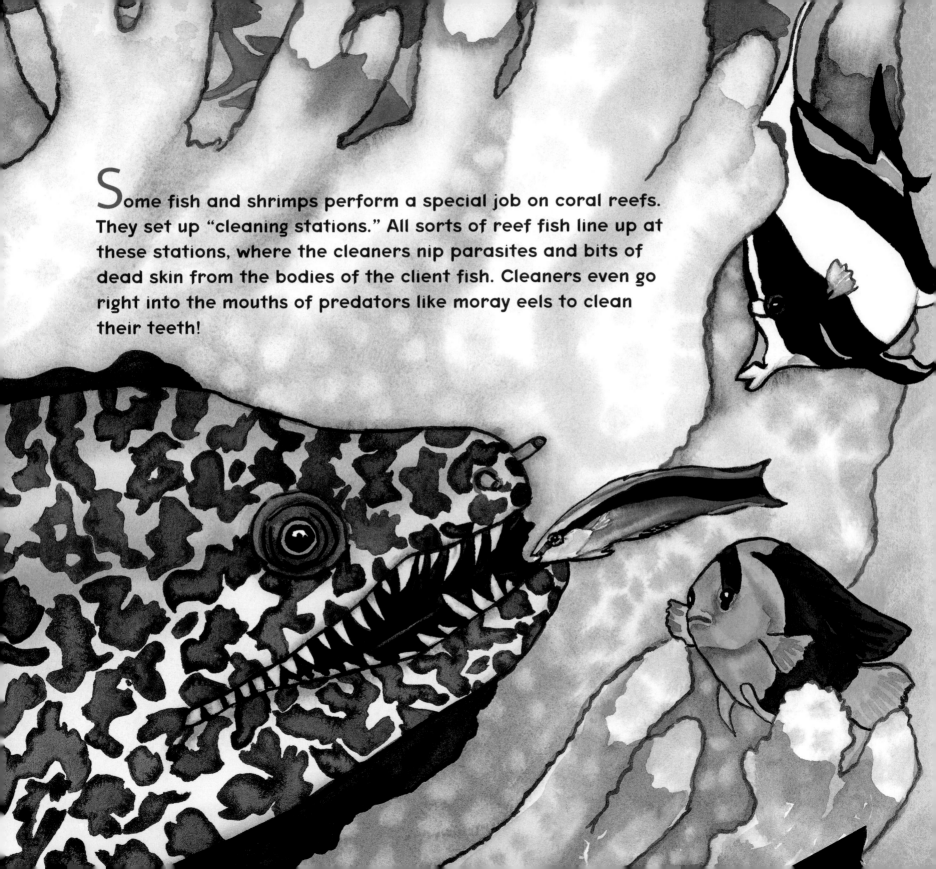

Some fish and shrimps perform a special job on coral reefs. They set up "cleaning stations." All sorts of reef fish line up at these stations, where the cleaners nip parasites and bits of dead skin from the bodies of the client fish. Cleaners even go right into the mouths of predators like moray eels to clean their teeth!

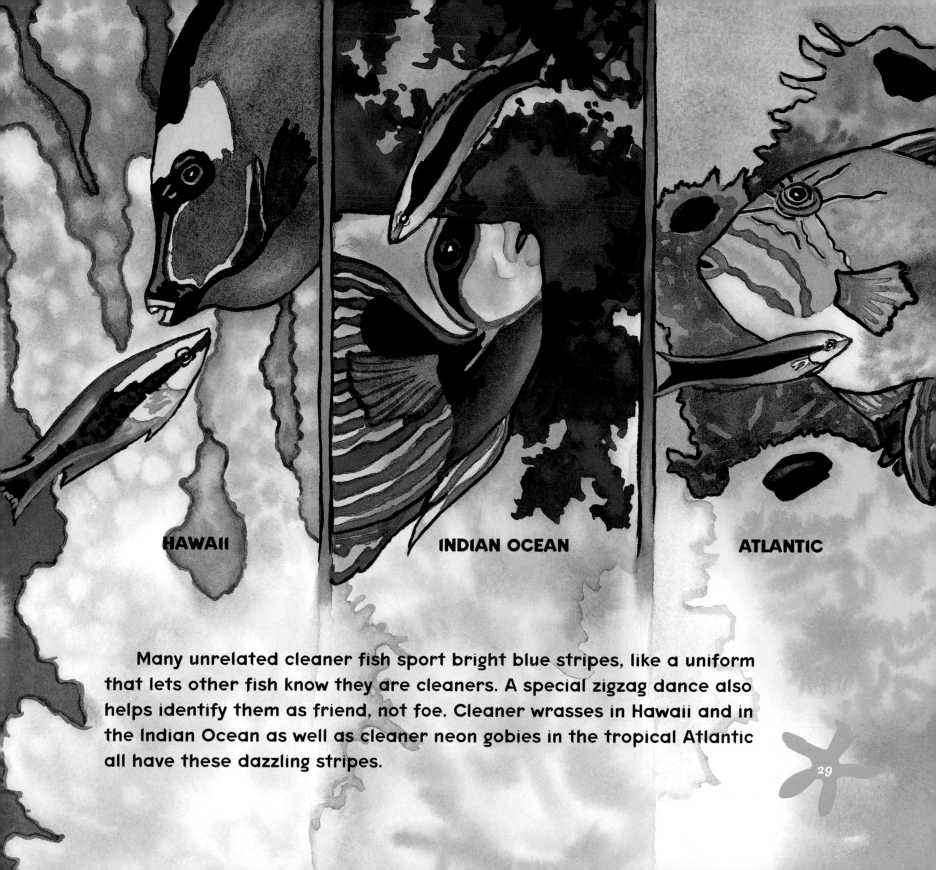

HAWAII

INDIAN OCEAN

ATLANTIC

Many unrelated cleaner fish sport bright blue stripes, like a uniform that lets other fish know they are cleaners. A special zigzag dance also helps identify them as friend, not foe. Cleaner wrasses in Hawaii and in the Indian Ocean as well as cleaner neon gobies in the tropical Atlantic all have these dazzling stripes.

Sometimes one kind of reef animal uses another as its home. One common relationship is between sea anemones and anemonefish. Sea anemones are related to corals and live in all the world's oceans. Each anemone looks much like a giant coral polyp, an inch or more in diameter, but without the limestone skeleton.

30

Anemonefish live right among the tentacles, yet they are not stung. By living there, the anemonefish are protected from predators that stay away from the painful stingers. The fish also lay their eggs under the tentacles, where they are safe. In return, the fish remove parasites from the anemones. As they swim, their fins create water currents that help whisk away wastes. The anemonefish are also very aggressive and drive away other fish that might feed on the anemone.

People depend on coral reefs in many ways. Reefs provide barriers against waves, protecting beaches and coastal buildings from dangerous storms.

A number of important food fish live on coral reefs, and thousands of people around the world catch fish there to sell. Tourists who come to snorkel and dive put millions of dollars into the economy of countries with coral reefs.

Reefs can provide important medicines. For example, AZT, a drug used to treat AIDS, comes from a sponge that lives on Caribbean reefs. Other drugs that are used to treat cancer, heart disease, and ulcers also come from reefs.

Reefs can benefit people, but those same people can damage reefs. Twenty percent of the world's coral reefs have disappeared in the last twenty years or so, mostly because of human activity. Many things can harm reefs. The anchors of boats bang into reefs, breaking them up. People who visit reefs can damage them by collecting coral souvenirs and by stepping on the coral.

Overfishing has reduced the numbers of some species to dangerously low levels. People sometimes use dynamite to stun fish, and the dynamite damages the reefs. They may use poison, which kills many kinds of living creatures. Reef fish are also collected to sell for home aquariums, sometimes in large numbers. In areas where lots of people live, rivers and sewers that empty into the ocean can carry pollution and choking mud onto reefs.

Changes in the modern world can be big threats to coral reefs. The crown-of-thorns sea star, which is as big as a dinner plate, eats living coral.

Every fifteen to twenty years, the number of these hungry creatures increases, and they kill large sections of reefs. Some scientists think pollution of the sea by fertilizer from farms helps feed the sea star larvae, increasing the numbers of these killers.

Global warming also threatens the world's reefs by warming up the ocean. When the water temperature gets too warm, the corals lose the algae that live in their cells. This is called coral bleaching. Often, the corals then die. In 1998, coral bleaching killed huge numbers of corals in the Indian Ocean as well as many along the Great Barrier Reef. Global warming also leads to higher sea levels and more powerful storms, both of which can endanger reefs.

MARINE
Preserve
DO NOT ENTER

People throughout the Tropics are trying to save coral reefs. They work at educating tourists so they will respect the reefs and treat them with care. They are setting up underwater preserves, where fish and other marine life can live and reproduce undisturbed by human activity.

Governments of some countries are working toward slowing global warming by reducing the amount of carbon dioxide and other greenhouse gases they release into the air.

Scientists are studying reef life, too. The more we understand about how coral reef plants and animals live, the easier it will be to make good decisions about how to manage reefs and preserve them for the future.

39

Index of Coral Reef Dwellers in this Book

(with scientific names)

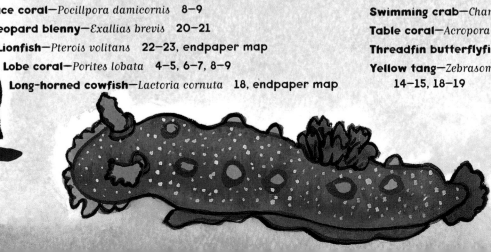